BEYOND THE ELECTRON

BEYOND THE ELECTRON

by

SIR J. J. THOMSON
O.M., F.R.S.

Master of Trinity College
Cambridge

A Lecture
given at Girton College on
3 *March* **1928**

CAMBRIDGE
AT THE UNIVERSITY PRESS
1928

CAMBRIDGE
UNIVERSITY PRESS

University Printing House, Cambridge CB2 8BS, United Kingdom

Cambridge University Press is part of the University of Cambridge.

It furthers the University's mission by disseminating knowledge in the pursuit of education, learning and research at the highest international levels of excellence.

www.cambridge.org
Information on this title: www.cambridge.org/9781316626146

© Cambridge University Press 1928

First published 1928
First paperback edition 2016

A catalogue record for this publication is available from the British Library

ISBN 978-1-316-62614-6 Paperback

NOTE

This Lecture, given at Girton College on March 3rd, 1928, was the first of a series of Founders' Memorial Lectures which the generosity of an old student has enabled the College to establish.

BEYOND THE ELECTRON

N OT so very long ago the atom was thought to be a terminus beyond which it was impossible from the nature of things to penetrate. The atom was regarded as indivisible, impenetrable, eternal, unaffected by heat, electricity or any other physical agent. The inside of the atom was regarded as a territory which the physicist could never enter. Then there came a time when the sanctuary of the atom was invaded, and it was found that the atom was built up of smaller parts—of electrons carrying a charge of negative, and of protons carrying a charge of positive electricity. Means were devised for counting the number of electrons in an atom, and it was found that the atom, instead of being just the little hard solid particle of the original view, was a very complex thing, comparable in complexity with the Solar System. It was found moreover that it was this complexity, this fine structure inside the atom, which endowed matter with its electrical and chemical properties. If we have any insight into these properties, it is due not so much to the idea that matter consists of a large number of small particles as to the knowledge we have obtained of individual particles and their electronic structure. Experiments told us what was the mass of the electron and the total charge of electricity associated with it; they did not however tell us anything about its structure; they did not tell us for example whether it was just a point charge of negative electricity or whether like the atom it was made up of smaller parts, sub-electrons and sub-protons as it were. It is true that there is no evidence that there are different kinds of electrons as there are different kinds of atoms, but this may be because one kind of electron is so much more stable than any other that the number of the latter is quite insignificant. In the absence of definite knowledge it was natural to begin with the simplest assumption and regard the electron as a single point charge surrounded by a structureless medium. The mathematics are simpler on this view than on any other; this however is not conclusive, as there is no evidence that the convenience of mathematicians has been a dominant factor in the scheme of the Universe.

It was therefore not improbable that in the light of further knowledge this view of the electron might prove as untenable as the corresponding view for the atom. My object this afternoon is to point out that this further knowledge has come, and that the electron and its surroundings must have a structure very different from that first assigned to them.

Perhaps some of you may ask, Is not going beyond the electron really going too far, ought one not to draw the line somewhere? It is the charm of Physics that there are no hard and fast boundaries, that each discovery is not a terminus but an avenue leading to country as yet unexplored, and that however long the science may exist there will still be an abundance of unsolved problems and no danger of unemployment for physicists.

The reason why we have to give up the old view of the electron is that it has recently been shown that a moving electron, even a uniformly moving one, is always accompanied by a series of waves. These waves as it were carry it along and determine the way it is to go; thus a moving electron is a much more complicated thing than a small point charge of electricity in uniform motion.

The clearest evidence for the existence of these waves round the electron is, I think, given by a research by my son, Professor G. P. Thomson, on the effects to be observed when electrons pass through exceedingly thin plates of metal—the metal has to be exceedingly thin, far thinner than the thinnest gold leaf. These plates, however, when they are obtained are exceedingly valuable physical instruments, for they enable us to test whether anything passing through them is a stream of particles or a train of waves. For suppose that we have a thin pencil of rays and we wish to determine whether these are a swarm of particles all moving in one direction or a train of waves. In either case if the pencil fell directly on a photographic plate it would produce a sharply defined image. Now let us see what would be the effect on this image of interposing the thin plate of metal. If the pencil consists of particles, these will strike against the molecules of the plate and be deflected by the collisions—how much each particle is deflected is within certain limits very much a matter of chance, so that some particles will be deflected more than others. Thus when they come out of the plate the particles will not all be moving in the same direction; the stream of particles will spread out into a cone; this will make the image they form on the photographic plate bigger and blurred, and it will become just a smudge without any definite pattern.

Suppose, however, that instead of a stream of particles we have a train of waves, then in consequence of the regular spacing of the molecules in the plate, the plate will act like a diffraction grating; and if the distance between the molecules is comparable with the length of the waves, we know from the properties of such gratings that when we interpose the plate in the way of the beam the original spot will not become just a smudge, but will be surrounded by a series of bright rings whose radii bear definite ratios to each other.

Now Fig. 1 represents the effect found by my son when he passed a stream of electrons through the plate. You see a well-developed series of rings,

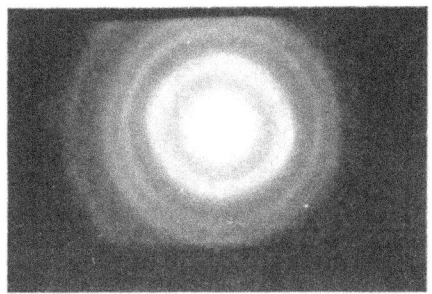

Fig. 1
Rings produced when electrons pass through
a thin plate of gold.

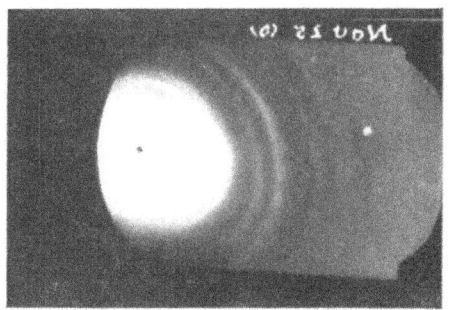

Fig. 2
Deflection of rings by a magnet.

and they are just in the position of the diffraction rings which would be produced if light of suitable wave length passed through the thin plate. That those rings marked the path of electrons was shown by bringing a magnet near the photographic plate; the rings were displaced just as the path of the electrons would be displaced (Fig. 2); this shows that the blackening of the plate is due to electrons and not to waves of light, for these would not have been affected by the magnet. Thus it appears that the electrons in their path through the metal are bent, not like particles would be bent, but in all respects like waves of a suitable wave length. Hence we conclude that the electron is accompanied by a train of waves, and that these waves have complete control over its path; the electron is compelled to follow the lead of the waves.

The thin metal film does more than detect the waves, it enables us to measure their wave length. My son did this and the results are most interesting, for it turns out that these electronic waves are of extraordinary high pitch; the pitch of the lowest is nearly a million times that of visible light, it is far higher than that of Röntgen rays, higher than all but the very hardest of the highest pitched rays hitherto known, the γ-rays emitted by radioactive substances. They introduce us to a new type of radiation whose properties may differ fundamentally from any type of radiation with which we are acquainted.

Just as it was found necessary to supplement the corpuscles which, on the old corpuscular theory constituted light, by systems of waves, so it turns out that bare corpuscles of electricity are insufficient to explain the properties of electrons, and that these, like the corpuscles of light, must be accompanied by systems of waves. This duality of corpuscles and waves seems to be in evidence in many regions of physics and may be of the nature of things.

To appreciate the importance of this we must consider for a moment how energy travels from one place to another. For example, when an electron changes its position its energy moves from the old position of the electron to the new: how does this energy travel? I may perhaps make the meaning of the question clearer if I take for a moment the old idea of the electron as a sphere 10^{-13} cm. in radius. When this moves does the energy travel, so to speak, as an inside passenger within the sphere, or is the energy located in the space outside the sphere, and so has to buffet its way through the ether when it moves from one place to another? If, as I do, we believe with Faraday and Clerk Maxwell that the properties of charged bodies are due to lines of force which spread out from them into the surrounding ether, we must place the energy of the electron in the space outside the little sphere which is supposed to represent the electron. On this view all energy is in the ether and must travel from one place to another as waves through the ether. This transmission of energy through the ether instead of through the more obvious channels, was first put in a clear and precise form by my old friend Professor John Henry Poynting. His views lead to results which, though I believe them to be absolutely sound, are somewhat startling. For example, I suppose that most of you take for granted that the energy for the electric lamps flows from the power station to the lamps inside the copper wire which connects the one with the other. On Poynting's view this is not so, the energy does not travel inside the wire but keeps coming into the wire sideways from the space around it. The function of the wire is not so much to carry the energy inside as to guide the path of energy travelling outside. The energy travels as waves through the ether outside the wire at a speed which does not depend materially upon the size of the wire or the material of which it is made.

It is generally recognised that the transmission of electrical energy is by waves through the ether: can we go further and say that energy of all kinds is transmitted in this way? It may quite well be that this is not really going further, for all energy may be of the same kind, located in the ether and having to travel through it. When we attempt to extend the wave transmission of energy from electrical energy to what seem other kinds of energy, we are met with what at first sight might seem an insuperable difficulty, which is this: All electrical waves, whatever their wave length, travel with the same speed, that of light, through the ether, and the energy they carry must therefore travel with this speed. But when you or I move from one place to another and carry our energy with us, even the youngest of us are left sadly far behind by light. We must face the fact that energy may travel at any speed, and show that this is not inconsistent with its transmission through the ether. I think we can do this, for though it is true that electrical waves, long and short, all travel through a structureless ether with the same velocity, yet, if the ether has mixed with it electrons or any bodies charged with electricity, the conditions are entirely changed. The charged bodies are set in motion by the electrical waves, and as a consequence emit electrical waves themselves; these secondary waves unite with the primary one, change its character, e.g. alter its wave length without altering its frequency, and thus alter its velocity.

There is in the upper parts of the atmosphere a region called the Heaviside layer which on almost a cosmical scale is an example of the effect we are considering.

The Heaviside layer—whose existence, by the way, makes very long distance wireless possible—

is a region at a great altitude where the gaseous pressure is very low, and where the radiation coming from outside is much more intense than it is at the surface of the earth where the greater part has been absorbed by the atmosphere. In the Heaviside layer the radiation is so intense that it is able to split up many of the molecules of air into electrons and positive ions, so that this region, like the one we are contemplating, contains a supply of positive and negative charges. The behaviour of this Heaviside layer towards electrical waves is entirely different from that of the space close to the ground where there are no free electrons. In the lower regions all the waves, whatever their wave length, travel with the velocity of light; in the Heaviside layer none travel so slowly as this, and no two waves travel with the same velocity, unless they have the same wave length; the longer the wave length the greater the velocity of the waves. The connection between the velocity of the waves and the wave length is illustrated in the graph *APL* in Fig. 3, where the velocity is measured horizontally, the wave length vertically *. I shall call a medium of this type a super-dispersive medium.

---

* The algebraical relation between the velocity V and the wave length λ is

$$V^2 = c^2 + B\lambda^2;$$

B is a quantity proportional to the number of electrons per unit volume.

The vertical dotted line in the figure represents the relation between the velocity and wave length for waves in a normal medium. You see that the effect of these electrical charges is to increase the rate at which the waves travel. You may think that this is not very promising for our purpose, which is to explain how energy can move more slowly

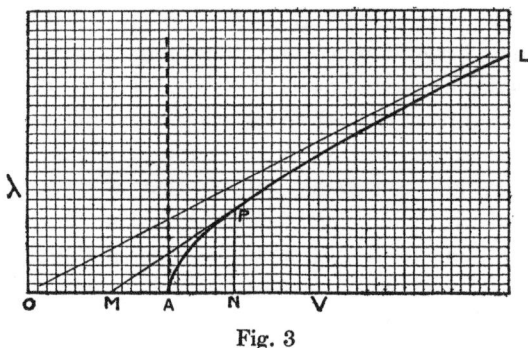

Fig. 3

than it does through the pure ether. In this you would be wrong, for it is one of the most important principles in connection with the transmission of energy by waves that we have to distinguish between the velocity of the waves and the velocity of the energy they are carrying; the greater the velocity of the waves the smaller is that of the energy. This fundamental principle is apt to be overlooked, for, in the most conspicuous cases of wave motion, sound and light, all the waves travel with the same velocity,

(16)

so that the question of the alteration in the speed of energy does not arise.

As this is a point of primary importance I must go into it a little more fully. When we excite waves, say water waves by throwing a stone into water or electrical waves by an electric spark, the initial disturbance is confined to a small area. When this is so the waves excited are not all of one wave length; to excite only one kind of wave the disturbance must be of a very special character and extend over a wide area. When as in practice the disturbance is very local we excite not a pure wave but a group of waves of different wave lengths, each of which in the super-dispersive medium will travel with a velocity different from that of the other waves. When the lengths of the waves excited, though not quite equal, all cluster around a particular value, the disturbance as it travels outwards will maintain its initial characteristics, it will in the main be concentrated within a space comparable with that within which it started; the energy will be concentrated in this space and move with the same velocity as the disturbance—this may be very different from that at which any of the waves move. The velocity at which the disturbance or the energy moves is called the group velocity; the velocity of the waves, the wave velocity. Let me illustrate the difference by a simple example. Take the case when there are two sets of waves, and let us represent one set by a procession of men, walking in a straight line at a constant speed and with a constant distance between each man and his nearest neighbour: the speed with which they walk represents the wave velocity, the distance between them the wave length. Let the other set of waves be represented by a procession of girls moving at a different speed from the men and separated from each other by a different distance; suppose the two processions are walking side by side. If the men represent the crests of one set of waves, the girls the crests of the other set, then when a man and a girl are together the crests of the two sets coincide and the disturbance and energy are a maximum at these places. Let us concentrate our attention on these places and find the velocity with which they travel. If an observer stands still at one place and waits long enough he will see a man and a girl side by side; but since the two processions are out of step the next man that passes will not pass at the same time as the next girl, and it may be a long time before he sees another couple. Could he get a richer harvest by walking forward, and if so, what is the pace at which he ought to walk? Let us call this pace W and suppose he starts walking when a couple is passing: how long will it be before the next girl overtakes him? If V is the pace at which the girls are walking and D the distance between them, he will have to wait for a time

$$\frac{D}{V - W};$$

and if v is the pace of the men and d the distance between them, he will have to wait for a time

$$\frac{d}{v-w}$$

for the next man. But if these times are equal, i.e. if

$$\frac{D}{V-W} = \frac{d}{v-w},$$

which will happen if $W = \dfrac{vD - Vd}{D-d}$, then when the girl appears there will be a man by her side, and if he continues at this pace the men and the girls will always pass him simultaneously and to him the procession will appear to consist of nothing but couples. This pace is the velocity with which the energy travels, and you see it may be entirely different from either V or v the wave velocities of the two components: in fact if vD were equal to Vd, the group velocity would be zero however great the wave velocities might be. We see from this analogy that though the velocity of the energy may lag far behind that of the waves, the path of the energy will be that of the waves; the waves guide the energy along the path it has to take. We can find the velocity of the energy in a very simple manner from the diagram (Fig. 4) of the connection between the velocity of the waves and their wave lengths.

A point P on this diagram represents a wave whose velocity is ON and wave length PN: the velocity with which the energy is carried by a group of waves whose wave lengths cluster around the value PN is OM, where M is the place where the tangent to the curve at P cuts the horizontal axis. It can be shown that $OM \cdot ON = c^2$, where c is the velocity of light in a non-dispersive medium; thus the faster the velocity of the waves the slower that of the energy, and we see from the curve that OM may, according to the

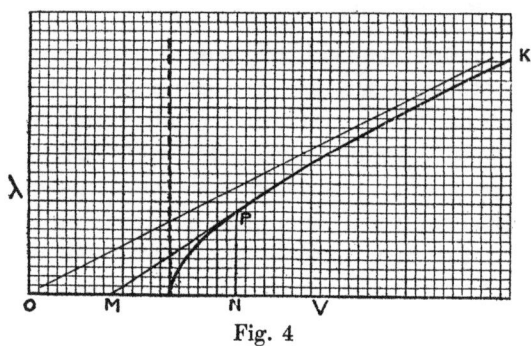

Fig. 4

position of P, have any value from zero up to the velocity of light. Thus we see that though energy may have to be transmitted by waves this does not imply that it must travel with the velocity of light. When the ether is mixed with charged bodies energy can pass through it at speeds varying from zero to the velocity of light, the velocity of the energy depending on the wave length of the waves which guide it.

A storm at sea often affords striking illustrations of the difference between the velocity of the energy and that of the waves. Waves in deep water, like electrical waves through ether mixed with charged particles, do not all travel at the same rate; the longer the water wave the greater its velocity, though the increase is not so great as for electrical waves. Now a stormy sea does not, as far as my experience goes, resemble the beautifully regular corrugated pattern which you see in some pictures; it is much more closely represented by great lumps of water spread in an untidy fashion over the surface; in other places there are waves, but they are much lower than the lumps. If you fix your attention on a crest of a wave near one of these lumps, you will see that the crest is travelling faster than the lump. The lump is a place where there is a great accumulation of energy; this energy travels with the lump, and as observation shows does so more slowly than the waves which accompany it.

I must direct your attention to one feature in the diagram for the electrical waves which is of primary importance. The wave length PN is equal to the velocity of the wave ON multiplied by the time of vibration, so that on the diagram the time of vibration is equal to PN/ON. Now you see that PN/ON never exceeds a certain limit, since P is always below the straight line OK. Thus the time of vibration of the waves must always be less than a finite quantity: this implies that only vibrations above a certain pitch can travel through the medium; it is opaque to all lower notes.

Thus we see that a medium which by itself can only transmit energy with the velocity of light may through the presence of electrical charges be put in a super-dispersive state and so modified that it is able to transmit energy at any speed less than that of light. The energy as it moves from one place to another is accompanied by a train of waves; these waves determine the path of the energy but, except when they pass through the particular region where the energy is concentrated, do not contain an appreciable amount of energy. The wave length of the waves varies with the speed of the energy, the greater the speed the shorter the wave length; any particular speed requires a special kind of wave to carry it. The speed of the energy is very much less than that of the waves; the product of this speed and wave velocity is equal to the square of the velocity of waves through pure non-dispersive ether; the energy system and the waves are in resonance. Now this is exactly the state of things which the experiments I have described show to exist in the electron; we have energy located at the electron itself, but moving along with it and guiding it, we have also a system of waves. In an appendix to this lecture I give a mathematical investigation of the passage of waves through a super-dispersive medium containing electrical charges; it follows from this that the relation between u the velocity of the electron and λ the wave length of the waves which accompany it is expressed by the equation

$$\frac{u\lambda}{\sqrt{1 - \dfrac{u^2}{c^2}}} = C,$$

where C is a constant and c the velocity of light, 3×10^{10} cm./sec. This is exactly the relation which my son finds to exist between u and λ in his experiments, and from these we can find the value of C; for when u was 10^{10} cm./sec. λ was $7{\cdot}8 \times 10^{-10}$ cm., hence $C = 8{\cdot}3$. The frequency of the waves, i.e. the number of vibrations per second, is equal to

$$c^2/C\sqrt{1 - \frac{u^2}{c^2}},$$

so that the smallest frequency is c^2/C, i.e.

$$1{\cdot}08 \times 10^{20}.$$

Thus the electron behaves as if it were within an atmosphere containing charges of electricity.

The view of the electron given above supposes that it has a dual structure, one part of this structure, that where the energy is located, being built up of a number of lines of electric force, while the other part is a train of waves in resonance with the electron and which determine the path along which it travels. This view of the electron has striking similarities with the view of the structure of light suggested by me in a paper in the *Philosophical Magazine*, Oct. 1924. On this view light, like the moving electron, has a dual structure, one part consists of a closed ring of electric force in which is located the energy of the light; the other part, as in the electron, is a system of electrical waves which do not carry energy themselves but determine the path of the carrier of the energy. These waves are in resonance with the ring, and the frequency of the light is proportional to the energy in the ring; thus the consequences of this structure for light are in accordance with Planck's law. If light possesses this dual structure the difficulty of reconciling the electrical properties of light, which seem to demand a corpuscular theory, with its optical properties such as the phenomena of interference which demand an undulatory theory, disappears; for the waves would show interference of the ordinary kind and would lead the energy into the bright and away from the dark parts of the interference pattern.

This duality is a necessary consequence of the transmission of energy through the ether by waves, for this involves two things, the transmission of energy and the propagation of the waves; this difference always exists, but it is obscured in the case of light waves because the velocity of transference of energy is in this case equal to the velocity of wave propagation. But this is, so to speak, an accident; the energy may, as we have seen, move very much more slowly than the waves, and then the duality is unmistakable.

If we confine our attention to optical phenomena, the waves are all-important and we do not have to consider anything else, hence we can be content with an undulatory theory: if we study the electrical properties we are concerned with the energy and may be content with a corpuscular one, where attention is concentrated on the carriers of the energy. If we concentrate on the waves we have an undulatory theory, if on the energy a corpuscular one. Thus both the undulatory and the corpuscular theory expressed a part but not the whole of the truth: in all optical phenomena, as well as those concerned with the movement of electrical charges such as cathode, a and β rays, both corpuscles and waves are present and must be taken into account.

The behaviour of the electron indicates that it is moving through a super-dispersive medium: the question arises, Is this medium confined to the immediate neighbourhood of the electron, as it might be if the electron like the atom was built up of smaller parts charged with electricity, or has the ether itself a structure of this kind? The ether is certainly not super-dispersive for ordinary light, but that does not preclude it from being super-dispersive for vibrations many thousand times more rapid, for dispersion is dependent upon rapidity of vibration. Thus glass shows no dispersion for the long electrical waves used in wireless telegraphy because the time of vibration of these is very long compared with the time of vibration of the molecules of the glass; it is dispersive for visible light because the time of vibration of the light is not very greatly different from that of the molecules. If the ether had a structure whose time of vibration was very much shorter than that of the vibrations of visible light but greater than the time of vibration of the electronic waves, it would be dispersive for these waves but not for light.

Let us consider how we could distinguish between these views: on the first view the super-dispersive region would be confined to the neighbourhood of the electron, the dimensions of this region would be those of the electron.

The value usually attributed to the diameter of the electron is 10^{-13} cm.; this is not the result of any direct measurement but is calculated from the known mass and charge of an electron. The possession of an electric charge endows a system with mass, the amount of which will depend upon the configuration of the system. Assuming that the electron consists of a point charge of electricity contained within a sphere of radius a and that the whole mass of the electron is due to its charge, it follows that if m_0 is the mass of a stationary electron and c the velocity of light,

$$m_0 c^2 = \frac{2}{3} \frac{e^2}{a};$$

since m_0, e, and c are known, a can be calculated.

It is in this way that the value 10^{-13} cm. has been arrived at. It is as we see dependent on the view we take of the structure of the electron, for no one has measured the diameter of an electron. If we took a different view of its structure; if, for instance, we supposed that the electric field is not uniformly distributed round the electron but done up into bundles, we should get a higher value for the diameter. On the view we have been taking the most important dimension of the electron is the size of the region round the electron which is endowed by it with super-dispersive properties; on this point we can hope to get evidence by direct experiments, such as those made by G. P. Thomson on the rings produced when the electrons pass through thin sheets of metal; the radii of the rings give the wave length of the waves associated with the electron. The rings arise from the interference of these waves which must be in the super-dispersive region round the electron; the train of these waves must include several wave lengths, otherwise we should not get pronounced interference. The wave lengths of the waves measured by G. P. Thomson were, for electrons whose speed was 10^{10} cm./sec., about $7 \cdot 8 \times 10^{-10}$ cm./sec.; hence we conclude that the diameter of the super-dispersive region must be at least 10^{-9} cm., a very large value compared with the 10^{-13} cm. of the ordinary theory. By using less rapid electrons the wave length can be increased, but when the wave length becomes greater than the diameter of the super-sensitive state the interference will be very much blurred: hence by determining the stage at which blurring sets in we can form an estimate of the diameter of the super-dispersive region.

There is evidence that the rings become much less distinct when the velocity of the electrons is reduced much below 10^{10} cm./sec. and the wave length therefore considerably longer than 10^{-9} cm.; we are not however in a position to say that this blurring cannot be due to other causes.

Since well-defined rings have been found with wave lengths of 10^{-9} cm.: the diameter of the super-dispersive region must be at least as great as this. This size is not so great but that it could be reconciled with the properties of the electron, for we must remember that electronic waves have remarkably high frequencies, so that the dispersive power is being excited by very rapid vibrations. When exposed to lower frequencies this region may not show super-dispersive powers but behave like the normal ether, and thus escape notice. It is thus not surprising that the results got from the study of electronic waves indicate a higher value for the size of the electron than other methods.

Among the γ rays given out by radio-active substances, there are some whose wave lengths are comparable with those of electronic waves; they are however not accompanied by electrons. If the

super-dispersive state is confined to the neighbourhood of the electron, the velocity of these γ rays
will not be the same as the velocity of the electronic
waves: if on the other hand the super-dispersive
property is a property of the ether generally, the
velocity of the γ rays will be the same as the velocity
of electronic waves of the same length, and therefore greater than the velocity of light. Now the
wave lengths of the γ rays can be measured by a
method similar in principle to that used to measure
the lengths of the electronic waves: the frequency
of the waves can also be measured by finding the
speed of the β particles expelled by the γ rays from
metals on which they fall. If it be assumed that all
the energy of the rays goes into the β particles,
E, the energy of the γ ray, is determined. The frequency ν follows from E by the relation $h\nu = E$.
The velocity of the ray is $\lambda\nu$. Measurements of λ
for very hard γ rays have been made by Kovarik
and measurements of the frequency by Ellis and
Skinner: it appears from these measurements that
the value of $\lambda\nu$ does not differ much, if at all, from
that of light, and hence that the super-dispersive
property is due to the presence of the electron, in
other words that the electron provides its own ether.
We should expect that when these hard γ rays came
near an electron the effect of the collisions would
be much greater than for those between electrons
and light of lower frequency. The medium would

not be super-dispersive for the lower frequencies, while it would be so for the hard γ rays. When however it is super-dispersive, the refractive index is much less than unity: for example, for waves of the frequency of those in G. P. Thomson's experiments, the refractive index was only $\frac{1}{3}$. Thus when the γ rays enter the region round the electron they will be passing into a less refracting medium and will therefore be bent as if they were repelled by the electron. Since the change in refractive index is very large the bending of the γ ray due to refraction may be very considerable. The deflection of γ rays involves a change in their momentum, and by the principle of the conservation of momentum, the momentum lost by the rays must be gained by the electrons. These very hard γ rays have masses comparable with that of an electron, so that the collision between a ray and an electron will be comparable with one between two bodies of not very different mass, one of which is at rest. In such a collision when one is bent through a considerable angle, there is a transference of a considerable fraction of the energy of the moving one to the one at rest. Thus we should expect the γ rays to lose energy by these collisions; this loss of energy lowers their frequency. The change in frequency by collision is what is known as the Compton effect, so that on this view this effect ought to be abnormally great with the γ rays.

On the view that the electron is a system built

up of many smaller parts, it might vibrate in many different ways and with different periods. The number of these periods might be very great indeed, but the different periods would form a series, the members of which are separated from each other by finite intervals. The electron might be able to vibrate in a great many different periods but not necessarily in any period arbitrarily chosen. Now there must be resonance between the waves accompanying the electron and the electron itself. The momentum of the electron and also its energy are connected by simple relations with the frequency of the guiding waves; thus it is shown in Appendix A that the energy of the electron is proportional to the frequency of these waves, and that the product of the momentum of the electron and the wave length is constant. It follows therefore that since the frequencies and the wave lengths do not range over all possible values, neither will the momentum nor the energy. The possible values of the momentum, for example, will be separated by finite intervals, so that the increase in momentum will not take place continuously but by jumps; this corresponds to a kind of quantization of the momentum. Thus on this view quantization of dynamical quantities is the result and expression of the structure of the electron; only such motions are possible, or at any rate stable, as are in resonance with the vibrations of the underworld of the electron.

On almost any view that could be taken of its structure, the electron would have definite periods of vibration. For when it is in equilibrium the distribution of the electric field around it must be such as to make the potential energy a minimum. If this distribution is disturbed, say by the passage through the field of a high speed cathode ray, the new distribution will not be in equilibrium but will oscillate about the normal one. The times of these oscillations will be comparable with the time taken by light to pass over a distance equal to the linear dimensions of the electron.

The electrons are deflected by electric and magnetic forces, but the electron follows the path of the waves and therefore the paths of the waves must be deflected by these forces. The paths of rays of light are however straight lines unless the refractive index of the medium changes from point to point, as it does for example in a mirage when the air next the ground is hotter and less refractive than that higher up. Thus if the electric and magnetic forces are to bend the path of the rays, they must do it by making the refractive index of the super-dispersive region round the electron vary from point to point. We should expect this on the view we have been considering, for the refractivity depends on the way the electrical systems are dispersed through this region. If the electron is far away from other charged bodies, these systems will be symmetrically arranged round the electron; there will be as many on one side as the other, and the refractive index will be the same on the two sides. If however a charged body is brought near the electron it will disturb the symmetry of the arrangement, as it will make the lines of force round the electron crowd into one side and leave the other. The refractive index at any part of the region around the electron will change with the number of lines of force passing through that part, hence the change in the distribution of the lines of force produced by the charged body will make the refractive index on one side of the electron different from that on the other. Thus the waves will be bent as they pass from one side of the electron to the other, and as the waves are bent the path of the electron will be bent too. The mathematical theory of this effect is given in Appendix B.

I have attempted in this lecture to show that the properties of the electron recently discovered lead to the view that the electron is not the final stage in the structure of matter but that it has itself a structure, being made up of smaller parts which carry charges of electricity: such a structure would give to it the property it has lately been found to possess. The results of this theory coincide in many respects with those that follow from the extremely interesting theory of wave mechanics which we owe to M. Louis de Broglie, and which has been extended by Schrödinger and others. The coincidences are remarkable because two theories could hardly be more different in their points of view. M. de Broglie's theory is purely analytical in form; the one I have brought before you this afternoon is essentially physical. I have tried to show that the recently discovered properties of the electron are of the same type as those we are already familiar with in other branches of Physics, and to picture a structure for the electron which would endow it with those properties.

It is interesting to find that with this conception of the electron the application of the methods of what is known as "Classical Dynamics" leads to some results which are supposed to be peculiar to "Quantum Mechanics," indicating I think that this type of mechanics may only be required when a particular view is taken of the nature of the electron.

The experiments I have described, as well as those made by Davisson and Kunsman and Davisson and Gerner, on the reflection of electrons by crystals, have opened up new fields for experiments which we may hope with confidence will throw much light on that fundamental question, What is the nature of the electron?

APPENDIX A

The equations for the transmission of waves through a medium containing electrical charges are as follows:

X, Y, Z are the components of the electric force, α, β, γ those of the magnetic; x_r, y_r, z_r the coordinates of an electric charge whose mass is m_r; c the velocity of light.

There are three equations of the type

$$\frac{dX}{dt} + 4\pi c^2 \Sigma e \frac{dx_r}{dt} = c^2 \left(\frac{d\beta}{dz} - \frac{d\gamma}{dy}\right) \ \ldots(1),$$

three of the type

$$\frac{dX}{dz} - \frac{dZ}{dx} = \frac{d\beta}{dt}\ldots\ldots\ldots\ldots(2),$$

and three for each charge of the type

$$m_r \frac{d^2 x_r}{dt^2} + n_r^2 x_r = Xe + \left(\beta \frac{dz_r}{dt} - \gamma \frac{dy_r}{dt}\right) e \ldots(3).$$

From these we get

$$\frac{d^2 X}{dt^2} + 4\pi c^2 \Sigma e \frac{d^2 x_r}{dt^2}$$

$$= c^2 \left(\frac{d^2 X}{dx^2} + \frac{d^2 X}{dy^2} + \frac{d^2 X}{dz^2}\right) - c^2 \frac{d\psi}{dx}\ldots\ldots(4),$$

where $\qquad \psi = \frac{dX}{dx} + \frac{dY}{dy} + \frac{dZ}{dz}.$

We may leave out the term in ψ as without significance in wave motion when the equations are linear.

Let us suppose that all the charged particles have the same mass, m', the same times of vibration and numerically equal charges; then if N is the number of electrical charges per unit volume,

$$4\pi c^2 \, \Sigma e \, \frac{d^2 x_r}{dt^2} = 4\pi N c^2 e \, \frac{d^2 x}{dt^2}.$$

In the super-dispersive state $m' \dfrac{d^2}{dt^2}$ is large compared with n^2 and with $(\alpha^2 + \beta^2 + \gamma^2) \, e^2/m'$, so that by (3) the right-hand side is equal to

$$4\pi c^2 \, (Ne^2/m') \, X$$

and (4) becomes

$$\frac{d^2 X}{dt^2} + 4\pi^2 B X = c^2 \left(\frac{d^2 X}{dx^2} + \frac{d^2 X}{dy^2} + \frac{d^2 X}{dz^2} \right) \ldots (5),$$

where $B = c^2 Ne^2/m'$.

This is the equation for wave motion in a super-dispersive region.

Take the case of a plane wave represented by

$$X = A \cos \frac{2\pi}{\lambda} (vt - z),$$

where v is the phase velocity and λ the wave length. We find from (5)

$$v^2 = B\lambda^2 + c^2 \ldots \ldots \ldots \ldots (6).$$

If u is the group velocity,

$$u = v - \lambda \frac{dv}{d\lambda} = \frac{c^2}{v},$$

hence $uv = c^2$; so that by (6)

$$c^4 = B\lambda^2 u^2 + c^2 u^2;$$

so that

$$\frac{\lambda u}{\sqrt{1 - \dfrac{u^2}{c^2}}} = \frac{c^2}{\sqrt{B}} \quad \ldots\ldots\ldots\ldots(7),$$

hence $\lambda u \Big/ \sqrt{1 - \dfrac{u^2}{c^2}}$ is constant, which is just the

relation between λ and u which holds in my son's experiments.

If ν is the frequency of the waves,

$$\nu = \frac{v}{\lambda};$$

and as $v = c^2/u$,

$$\nu = \frac{c^2}{\lambda u} = \frac{\sqrt{B}}{\sqrt{1 - \dfrac{u^2}{c^2}}} \quad \ldots\ldots\ldots\ldots(8).$$

Thus the smallest possible period for these waves is $\sqrt{B}$.

My son found that when the velocity of the electron was 10^{10} cm./sec. the wave length λ was $7\cdot8 \times 10^{-10}$ cm. Putting $u = 10^{10}$, $\lambda = 7\cdot8 \times 10^{-10}$ in equation (7) we find $\sqrt{B} = 1\cdot08 \times 10^{20}$, so that the smallest frequency of the electronic waves is $1{\cdot}08 \times 10^{20}$; this frequency corresponds to a wave length in air of $2{\cdot}7 \times 10^{-10}$ cm. This is far less than that of the hardest Röntgen rays and indeed of all but the very hardest γ rays, and is very small compared to the radius of an atom.

Since
$$\frac{1}{\sqrt{1-\dfrac{u^2}{c^2}}} = \frac{m}{m_0},$$

where m is the mass of an electron moving with the velocity u, and m_0 the mass when it is at rest, we see from (8) that

$$\nu = mc^2 \frac{\sqrt{B}}{m_0 c^2} \quad \ldots\ldots\ldots\ldots(9).$$

Now mc^2 is the total energy of the electron, hence we see that the frequency of the waves is equal to a constant multiplied by the energy of the electron. This is what is known as the quantum relation, and it is to be noticed that it appears not as a postulate from quantum mechanics but as a necessary consequence of the view that the energy of the electron is moving through a super-dispersive medium.

Another interesting point is that if we calculate by quantum mechanics the mass of a quantum of light of any period, the result is the mass of the electron which on our theory is associated with electronic waves of that period.

If μ is the refractive index,

$$\mu = \frac{c}{v} = \frac{u}{c},$$

but from (8)

$$\frac{v}{v_0} = \frac{1}{\sqrt{1 - \dfrac{u^2}{c^2}}},$$

where $v_0 = \sqrt{B}$; thus v_0 is the smallest value of v. Hence

$$\mu = \frac{u}{c} = \sqrt{1 - \frac{v_0^2}{v^2}} \quad \text{.........(10)}.$$

If $\qquad X = A \cos (pt - mz),$

then from (2)

$$\beta = \frac{m}{p} A \cos (pt - mz) = \frac{1}{v} A \cos (pt - mz) \quad (11).$$

The energy per unit volume due to the electric force is

$$\frac{1}{8\pi c^2} A^2 \cos^2 (pt - mz),$$

the energy per unit volume due to the magnetic force is

$$\frac{1}{8\pi} \beta^2 = \frac{1}{8\pi v^2} A^2 \cos^2 (pt - mz).$$

Since in a super-dispersive medium v is greater than c, the magnetic energy is not equal to the electrostatic as it is in non-dispersive media but is always smaller, and vanishes for waves which have the limiting frequency v_0 ; the difference between the electrostatic and magnetic energy is on the average equal to the kinetic energy possessed by the charges which give rise to the dispersive quality.

$$\text{For} \qquad m_1 \frac{d^2 x}{dt^2} = Xe$$
$$= eA \cos(pt - mz),$$
$$m_1 \frac{dx}{dt} = \frac{eA}{p} \sin(pt - mz),$$

and therefore the kinetic energy of these particles

$$\tfrac{1}{2} N m_1 \left(\frac{dx}{dt}\right)^2$$

is equal to

$$\tfrac{1}{2} \frac{Ne^2}{m'p^2} A^2 \sin^2(pt - mz).$$

But from (6)

$$\frac{Ne^2}{m'p^2} = \frac{1}{4\pi} \left(\frac{1}{c^2} - \frac{1}{v^2}\right),$$

hence the kinetic energy is equal to

$$\frac{1}{8\pi} A^2 \left(\frac{1}{c^2} - \frac{1}{v^2}\right) \sin^2(pt - mz);$$

the mean value of this is

$$\frac{A^2}{16\pi} \left(\frac{1}{c^2} - \frac{1}{v^2}\right),$$

which is equal to the difference between the mean values of the electrostatic and magnetic energy.

Thus the mean value of the total energy per unit volume is twice the mean value of the electrostatic energy and is therefore equal to

$$\frac{A^2}{8\pi c^2}.$$

The flow of energy across unit area per unit time is equal to the Poynting vector

$$\frac{1}{4\pi} X\beta = \frac{1}{4\pi} \frac{X^2}{v} \text{ by (11)}$$

$$= \frac{1}{4\pi} \frac{A^2}{v} \cos^2(pt - mz);$$

the mean value of this is equal to

$$\frac{A^2}{8\pi v}.$$

But if u is the velocity of the energy this flow must be equal to u times the mean energy, hence

$$\frac{A^2}{8\pi v} = u \frac{A^2}{8\pi c^2},$$

or $uv = c^2$.

By this method the velocity of the energy is deduced without introducing the idea of interference between trains of waves.

Since $\qquad \beta = \dfrac{X}{v},$

$$\beta = \frac{uX}{c^2},$$

or the magnetic force is proportional to the velocity of the energy.

If ν_0 is the limiting frequency, equation (5) may be written as

$$\frac{d^2X}{dt^2} + 4\pi^2\nu_0{}^2 X = c^2\left(\frac{d^2X}{dx^2} + \frac{d^2X}{dy^2} + \frac{d^2X}{dz^2}\right);$$

(41)

if ν is the frequency of the vibrations which are being transmitted,

$$\frac{d^2X}{dt^2} = -4\pi^2\nu^2 X;$$

hence the equation becomes

$$4\pi^2\left(\nu_0{}^2 - \nu^2\right)X = c^2\left(\frac{d^2X}{dx^2} + \frac{d^2X}{dy^2} + \frac{d^2X}{dz^2}\right),$$

or by (10)

$$c^2\left(\frac{d^2X}{dx^2} + \frac{d^2X}{dy^2} + \frac{d^2X}{dz^2}\right) + 4\pi^2\mu^2\nu^2 X = 0.$$

APPENDIX B

THE PATH OF AN ELECTRON UNDER THE ACTION OF AN EXTERNAL FORCE

The equations giving the path of a ray of light through a medium of variable refractive index μ are

$$\frac{d}{ds}\left(\mu\frac{dx}{ds}\right) = \frac{d\mu}{dx},$$

$$\frac{d}{ds}\left(\mu\frac{dy}{ds}\right) = \frac{d\mu}{dy},$$

$$\frac{d}{ds}\left(\mu\frac{dz}{ds}\right) = \frac{d\mu}{dz}.$$

If q is the velocity of the energy of an electron, u, v, w its components, $\mu = \dfrac{q}{c}$, so that these equations become

$$\frac{d}{ds}\left(\frac{u}{c}\right) = \frac{d\mu}{dx};$$

(42)

if ds is an element of the path of the energy $ds = q\,dt$, hence

$$\frac{du}{dt} = cq\,\frac{d\mu}{dx} = \tfrac{1}{2}\,c^2\,\frac{d\mu^2}{dx},$$

with similar equations for dv/dt, dw/dt.

Hence the path of the energy will be the same as that of a particle of mass m acted upon by a force whose components are $\tfrac{1}{2}\,mc^2\dfrac{d\mu^2}{dx}$, $\tfrac{1}{2}\,mc^2\dfrac{d\mu^2}{dy}$, $\tfrac{1}{2}\,mc^2\dfrac{d\mu^2}{dz}$; so that if these components can be represented as differential coefficients of a potential V, then in the super-dispersive region

$$\mu^2 = \mu_0{}^2 + 2V/mc^2,$$

where μ_0 is the value of μ when no forces act upon the electron.

For EU product safety concerns, contact us at Calle de José Abascal, 56–1°,
28003 Madrid, Spain or eugpsr@cambridge.org.

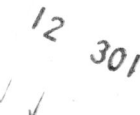